Marsha McCloskey's

Guide to

Rotary Cutting

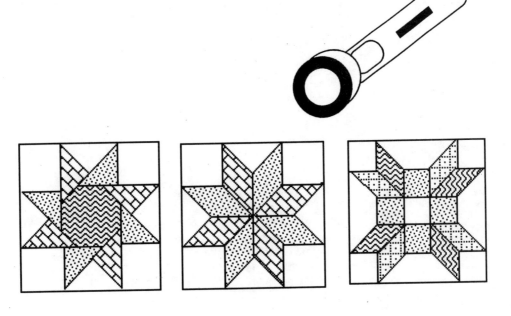

Ready reference for cutting
patchwork pieces without templates.

GUIDE TO ROTARY CUTTING

Also available from Feathered Star Productions:

PIECED BORDERS: THE COMPLETE RESOURCE
by Judy Martin and Marsha McCloskey, $19.95
FEATHERED STAR SAMPLER
by Marsha McCloskey, $7.95
FEATHERED STAR QUILTS
by Marsha McCloskey, $22.95
100 PIECED PATTERNS FOR 8" QUILT BLOCKS
by Marsha McCloskey, $11.95

GUIDE TO ROTARY CUTTING
© 1990 by Marsha McCloskey. All rights reserved.
Revised 1993

Feathered Star Productions
2151 7th Avenue West
Seattle, WA 98119
Phone/Fax (206)283-5214

Cover photography: Michael Craft

Printed in the USA
Eighth printing

No portion of this book may be reproduced or used in any form or by any means without prior permission of the author and publisher.

ISBN 0-9635422-1-4

Marsha McCloskey's

Guide to Rotary Cutting

Contents

Tools and Supplies	3
Choosing a Pattern	4
Making a Cutting Guide	4
Straight Grain and Bias	6
Cutting	7
Fabric Preparation	7
Straight Strips	7
Squares and Rectangles	8
Paper Templates for Odd-Size Shapes	9
Triangles	10
Half-Square Triangles	10
Quarter-Square Triangles	11
Long Triangles	12
Diamonds and Parallelograms	12
Octagons	13
Trapezoids	14
Other Shapes	14
Trimming Points for Easy Matching	15
Strip Piecing	16
Straight-Strip Piecing	16
Bias-Strip Piecing	17
Half-Square Units	18
Waste Triangles	19
Quarter-Square Units	20

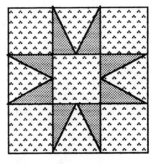

ABOUT THIS BOOKLET

Marsha McCloskey's GUIDE TO ROTARY CUTTING is a short and concise, ready reference for cutting patchwork pieces for quilts without templates. It includes measurements, rules and strategies for cutting squares, rectangles, triangles and other commonly used shapes. Also included are straight and bias strip piecing techniques, and some drafting - all the nuts and bolts information quilters need to look up occasionally as they are working. It will prove especially valuable in classroom situations when students, new to these techniques, need a written reference close at hand.

Caution!!

Rotary cutter blades are very sharp.

Be careful.

Keep the safety shield on when not in use.

Always roll the cutter away from yourself.

Keep this and other sharp tools out of the reach of children.

GUIDE TO ROTARY CUTTING

TOOLS AND SUPPLIES

Graph Paper: For rotary Cutting Guides, purchase large sheets (17" x 22") with 1/8" grid and heavy lines at the 1" increments.

Pencils: Colored pencils and a fine-tipped mechanical drawing pencil (an ordinary #2 lead pencil and a good sharpener will do).

Drawing Ruler: A C-Thru™ B-85 ruler, 2" x 18" with a red grid of 1/8" squares, for drafting your Cutting Guide.

Removable Tape: Removable tape holds tracing paper in place while you work and will not harm paper when it is removed. Use it also to tape paper templates to cutting rulers.

Scissors: A pair for cutting paper.

Rotary Cutter: These usually come in two sizes. I use the larger one with the 2" diameter blade. **The blades are very sharp, so take care not to cut yourself.** Keep a fresh refill blade on hand.

Cutting Mat: Made of various plastic materials, these mats come in several sizes and serve to protect your table and keep cutting blades sharp. My favorite mat measures 24" x 36" and covers half of my work table.

Cutting Rulers: Rulers for rotary cutting are 1/8" thick transparent Plexiglass™ and come in an amazing array of sizes and markings. These are the four I use the most:
1. A 6" x 24" ruler for cutting long strips. It is marked in 1", 1/4", and 1/8" increments (that's important!), with both 45° and 60° angle lines.
2. A 15" square for cutting large squares. It is marked in 1", 1/4", and 1/8" increments and is extremely useful for cutting large squares.
3. A 3" x 18" ruler for shorter cuts and medium-sized pieces where the previous two rulers prove too big.
4. A BIAS SQUARE™. This handy 8" square is marked in 1/8" increments with a 45° angle line running diagonally corner to corner. Invented for rotary cutting pieced half-square units, this tool has many uses.

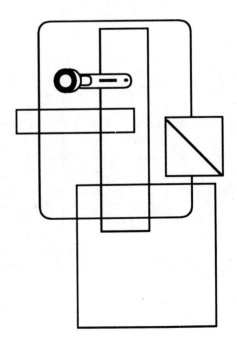

3

GUIDE TO ROTARY CUTTING

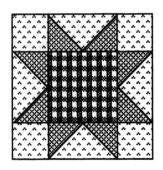

CHOOSING A PATTERN

There are thousands of pieced patchwork designs that are appropriate for rotary cutting. Look for patterns with simple, straight-sided shapes that have cutting dimensions that coincide with markings on the cutting rulers.

Squares, rectangles and right isosceles triangles are the easiest shapes to cut, but there are many ingenious ways to cut other shapes as well. Even shapes that have unmeasurable dimensions can be rotary cut by taping paper templates to cutting rulers to find proper measurements.

MAKING A CUTTING GUIDE

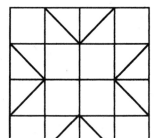

To get started you will need to make a Cutting Guide for your design. A Cutting Guide is a standard drafting of a block design on 1/8" graph paper. It clearly shows finished and cutting sizes of all the shapes needed. It also indicates grain lines, trimming points for matching and quick cut icons. The example shown is for a Sawtooth Star that measures 8" finished.

1. Draw a square the finished size of the design square on 1/8" graph paper. Ours is 8".

2. Study the design and determine the type of drafting needed. If it is a pattern on a grid; is it a nine-square, sixteen-square, twenty-five-square grid or something else. Our Sawtooth Star design is based on a sixteen-square grid. To find the size of each square in the grid, divide the measurement of the side of the square by the number of equal divisions required to make the grid. We divide our 8" side by 4 to equal 2". Draw the 2" grid squares inside the larger square.

3. Subdivide the grid squares to create the design. Note that not every line in the drafting will be a seam line. Some lines need to be dropped out to simplify piecing.

GUIDE TO ROTARY CUTTING

CUTTING GUIDE

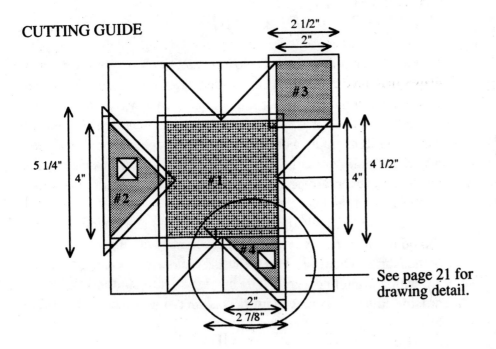

See page 21 for drawing detail.

4. Identify those shapes that need to be templates by coloring them in with colored pencils. The Sawtooth Star has four shapes: the large center square (#1), the large side triangle (#2), the small corner square (#3), and the small triangle (#4) that forms the star point.

5. Using the C-Thru™ ruler and the graph paper to guide you, add 1/4" seam allowances around each identified shape on the drawing. The seam allowance lines may overlap, but that's okay because none of these shapes will be cut out of the paper.

6. Read the discussion of Trimming Points for Easy Matching on page 15. Indicate trimming lines on your Cutting Guide.

7. Study the discussion of Straight Grain and Bias on page 6. Place the proper grain lines on each numbered shape on your Cutting Guide.

8. Study the directions for cutting half-square and quarter-square triangles on pages 10 and 11. Draw a half-square Quick Cut Icon on the #4 triangle and a quarter-square Quick Cut Icon on the #2 triangle.

QUICK CUT ICONS

Quick Cut Icons are little pictures that remind you how a triangle is cut.

Half-square triangles are made from a square cut in half.

Quarter-square triangles are made from a square cut in quarters.

GUIDE TO ROTARY CUTTING

STRAIGHT GRAIN AND BIAS

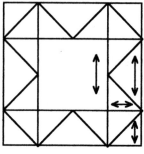

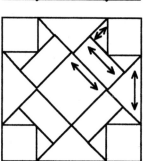

Fabric is made of threads woven together. Threads that run the length of the fabric, parallel to the selvage, are lengthwise straight grain. Threads that run across the fabric are crosswise straight grain. All other grains are considered bias. True bias runs at a 45° angle to the two straight grains. For the small pieces in patchwork both kinds of straight grain are considered equal. Long strips for borders and lattices, however, are best cut from the lengthwise grain as it is the more stable of the two.

Bias stretches and straight grain holds its shape, so fabric pieces should be cut with one or more edges aligned with the straight grain of the fabric. When you make a Cutting Guide, mark each shape to be cut with a grain line arrow. **The straight grain should fall on the outside edge of any pieced unit.** This applies to pieced units, design blocks, set pieces in the larger quilt, and the edge pieces in pieced borders.

The exception is that sometimes, in order to use a special print (such as a stripe) in a design the way you want to, a bias cut edge must fall on the outside edge of the block. On such a piece, a line of stay stitching 1/8" from the cut bias edge will stabilize the fabric and keep it from stretching.

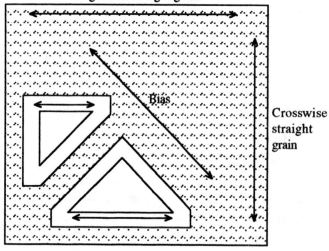

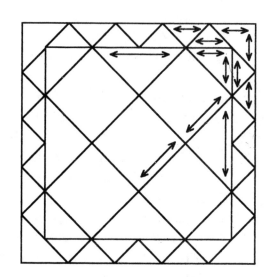

GUIDE TO ROTARY CUTTING

CUTTING

Fabric Preparation

For best results, choose lightweight, closely woven, 100% cotton fabrics. **The fabric you buy should be preshrunk, color tested, and ironed before it is cut.**

Wash lights and darks separately with detergent and warm water. Dry them in the dryer. If you suspect the dark colors might run, rinse those fabrics repeatedly in clear water until dye loss stops.

Straight Strips

In quilt construction, long strips of fabric are used for borders, and shorter ones are used for lattices. Narrow strips are used in some design blocks, and **the rotary method of cutting squares and rectangles begins with cutting strips of fabric.**

All strips are cut with the 1/4" seam allowance included.

1. Fold the fabric selvage to selvage, aligning the cross and straight grains as best you can. Place fabric on the rotary cutting mat with the folded edge closest to your body. Align the BIAS SQUARE™ with the fold of the fabric and place a cutting ruler to the left. When making all cuts, fabric should be placed to your right. [Note: If you are left handed, reverse the directions.]

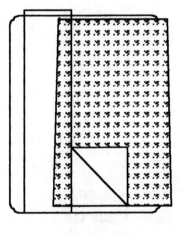

2. Remove the BIAS SQUARE™ and make a rotary cut along the right side of the ruler. Hold ruler down with left hand, placing the smallest finger off the ruler to serve as an anchor and prevent slipping. Stand comfortably with your head and body centered on the cutting line. Move hand along ruler as you make the cut, making sure the markings remain accurate. Use a firm even pressure as you cut. Begin rolling the cutter before you reach the fabric edge and continue across. **For safety, roll the cutter away from you. The blade is very sharp, so be careful!**

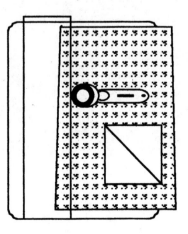

GUIDE TO ROTARY CUTTING

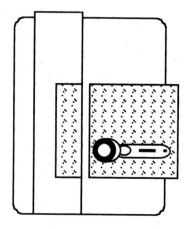

3. Fold fabric again so that you will be cutting four layers at a time. Cut strips of fabric the desired width as shown. Open fabric periodically to make sure you are making straight cuts. If fabric strips are not straight, use BIAS SQUARE™ to realign as in step #2.

SQUARES AND RECTANGLES

Strip Method

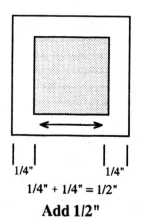

1/4" 1/4"
1/4" + 1/4" = 1/2"
Add 1/2"

1. First cut fabric in strips the finished measurement of the square plus seam allowances.

2. Using the BIAS SQUARE™, align the top and bottom edge of strip and cut fabric into squares the width of the strip.

3. Cut rectangles in the same manner, first cutting strips the length of the rectangle plus seam allowances, then cutting to the proper width.

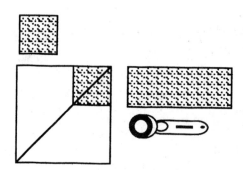

GUIDE TO ROTARY CUTTING

BIAS SQUARE™ Method

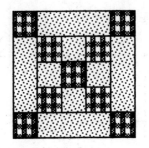

The strip method of rotary cutting squares yields a lot of squares of one fabric in a hurry. To cut just a few squares of a fabric, use the BIAS SQUARE™. One to six layers can be cut at one time.

1. Position the BIAS SQUARE™ on a corner of the fabric. Make two cuts along the edges of the ruler to separate the square from the rest of the fabric. The measurement used should be slightly larger than the final cut size of the square.

2. Turn the separated square around and position the BIAS SQUARE™ at the desired cutting measurement. Make the final two cuts to true up the fabric square.

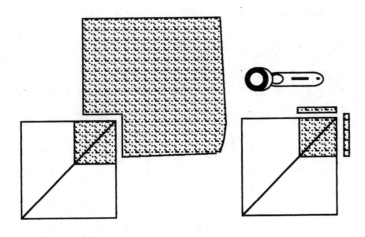

Paper Templates for Odd-size Shapes. Often shapes and cut dimensions of pattern pieces do not correspond with markings on standard cutting rulers. Just because a square measures 4 5/16" does not mean it cannot be cut with a rotary cutter. To cut odd-size squares, rectangles, diamonds, parallelograms, etc., make an accurate paper template of the shape (including the 1/4" seam allowance) based on your Cutting Guide. Carefully trace the shape on tracing or typing paper, cut it out and and tape it to the bottom of your cutting ruler with removable tape. You will then have the proper guide for cutting your shape.

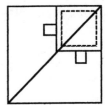

GUIDE TO ROTARY CUTTING

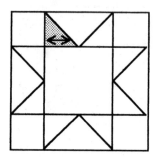

Triangles

Half-Square Triangles. To cut these triangles, cut a square and then cut it in half diagonally. The resulting triangles will have short sides on the straight grain of the fabric and the long side on the bias.

To allow for seam allowances, **cut the square 7/8" larger than the finished measurement of the short side of the triangle.**

1. Cut a square using the finished measurement of the short side of the triangle plus 7/8". (Or, just measure the short side of the triangle in your Cutting Guide, including seam allowances, from corner to tip to arrive at the proper size for the square.)

2. Take a stack of squares and cut diagonally corner to corner. Check the first triangles cut against the proper shape in your cutting guide to make sure they are the right size.

3. Use the BIAS SQUARE™ ruler to trim points for easy matching. (See page 15).

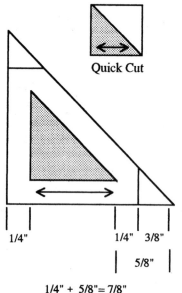

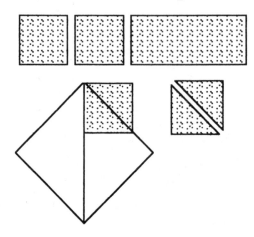

QUICK CUT ICONS

Quick Cut Icons are little pictures that remind you how a triangle is cut.

Half-square triangles are made from a square cut in half.

Quarter-square triangles are made from a square cut in quarters.

GUIDE TO ROTARY CUTTING

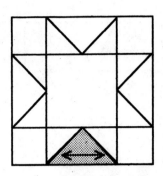

Quarter-Square Triangles. To cut these triangles, cut a square then cut it in half diagonally twice. The resulting triangles will have the long side on the straight grain and the short sides on the bias.

To allow for seams, **cut the square 1 1/4" larger than the finished measurement of the long side of the triangle.**

Quick Cut

1. Cut a square the finished measurement of the long side of the triangle plus 1 1/4". (Or, just measure the long side of the triangle in your Cutting Guide, including seam allowances, from tip to tip to arrive at the proper size for the square.)

2. Cut the square diagonally corner to corner. Without moving the resulting triangles, line up the cutting guide and make another diagonal cut in the opposite direction. Each square will yield four quarter-square triangles.

3. Use a template or ruler to trim 3/8" points off these triangles for easy matching. (See page 15.)

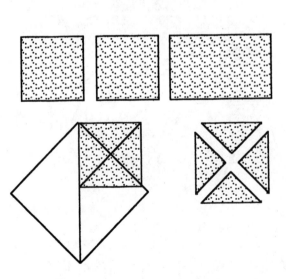

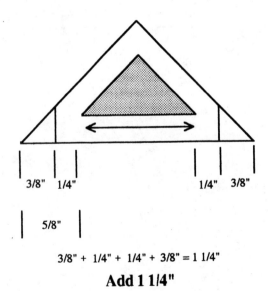

3/8" + 1/4" + 1/4" + 3/8" = 1 1/4"

Add 1 1/4"

GUIDE TO ROTARY CUTTING

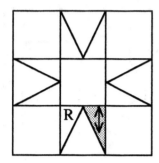

Long Triangles. The long triangles that form the points of the Unknown Star at the left can be rotary cut from rectangles. Note that the triangles must be cut as reversals or mirror images.

1. Cut straight strips as wide as the short side of the triangle, including seam allowances. The easiest way to arrive at this dimension is to draw the shape on graph paper and add 1/4" seam allowances.

2. Cut rectangles from strips, using the dimension of the long straight side of the triangle including seam allowances.

3. Cut rectangles from corner to opposite corner to yield two identical long triangles. To get reversed long triangles, cut the next rectangle diagonally in the opposite direction, or simply layer the fabrics, wrong sides together when the rectangles are cut.

4. Use a paper template to trim points for easy matching.

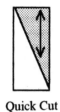

Quick Cut

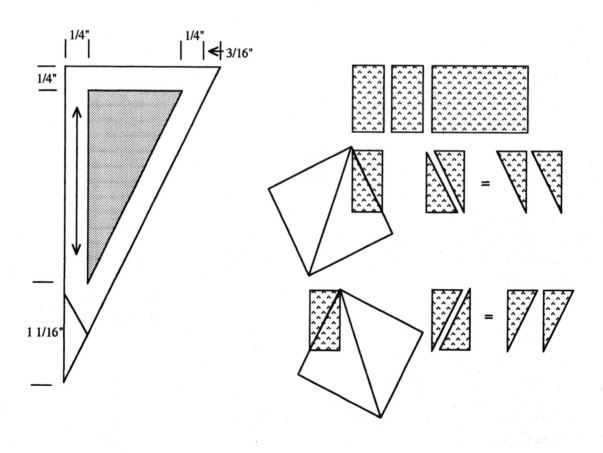

12

GUIDE TO ROTARY CUTTING

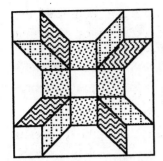
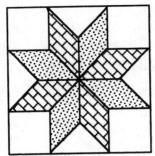
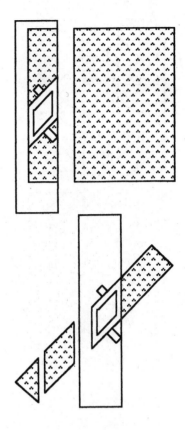

Diamonds and Parallelograms. Often these shapes have dimensions that don't match the markings on cutting rulers, so I make paper templates and tape them to the ruler to make measuring easy. If the dimensions do match the ruler, the template step is unnecessary.

1. Cut fabric strips the width of the finished diamond or parallelogram plus seam allowances. Determine this width by measuring the shape on your Cutting Guide or by making a paper template of the shape and taping it to the ruler.

2. Using the 45° angle line on the ruler, make a first cut on the fabric strip. Then, using either a measurement or a template, make successive diagonal cuts to form diamonds or parallelograms.

3. Use a template to trim points for easy matching or simply leave the points on; it will depend on the pattern.

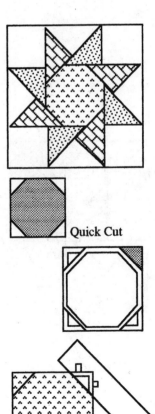

Quick Cut

Octagons. Octagons are squares with the corners cut off. To determine the dimension of the triangle-shaped corner pieces, study the drawn shape on your Cutting Guide. Complete the seam allowances around the square and the octagon. The triangle to be cut off is the difference between the two shapes with the seam allowances taken into consideration. Make a paper template of the triangle and tape it to the bottom of your ruler.

1. Cut a square the width of the octagon including seam allowances.

2. Make a paper triangle template as described above and tape it to the bottom of your ruler. Position the ruler as shown and trim off the four corners. Compare the cut shape to your Cutting Guide for accuracy.

GUIDE TO ROTARY CUTTING

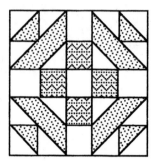

Quick Cut

Trapezoids. Trapezoids are triangles with the tops cut off. The shape cut off is a triangle. To determine the size of the triangle, study the shape on your Cutting Guide. Complete the seam allowance around the trapezoid and the triangle on which it is based. The difference between the two is the area to be trimmed off. Make a paper template of the triangle and tape it to the bottom of your cutting ruler.

1. Cut half-square triangles if the straight grain on the trapezoid needs to be on the short side; cut quarter-square triangles if the straight grain is on the long side.

2. With a paper triangle taped to the bottom of the cutting ruler, trim the smaller triangle from the larger one to make the trapezoid.

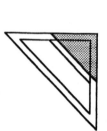 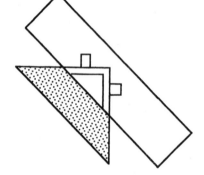

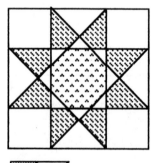

Quick Cut

Other shapes. Kites, equilateral triangles, etc. or any shape with straight sides can be cut with the rotary cutting techniques described here. Make a drafting or Cutting Guide for your block design as described on page 4. Base your template and cutting dimensions on the shapes in the Cutting Guide. Make paper templates for shapes and dimensions that do not coincide with ruler measurements. Play around with it and you'll be able to figure out how to cut any shape.

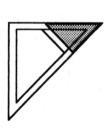 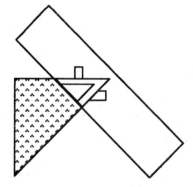

GUIDE TO ROTARY CUTTING

TRIMMING POINTS FOR EASY MATCHING

The purpose of trimming points is to take the guesswork out of matching cut patches before sewing. The general rule is to trim points of less than 90° at the 1/4" seam allowance lines, so they match neighboring pieces. Most shapes with 45° angles can be trimmed at the 1/4" line on the graph paper. Shapes with other angles are harder. Study the examples given here; notice that each shape is trimmed to fit the piece it will be sewn next to in the design. If you are in doubt as to which points to trim where, make paper templates with the points left on. Then simply match the seam lines of the two pieces to be sewn together with positioning pins and trim away any points that stick out.

Trim the points on rotary cut pieces as described below or make paper templates of shapes as guides.

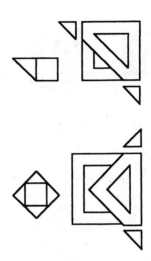

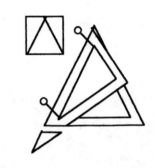

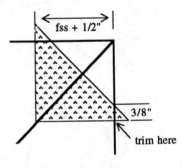

1. Half-Square Triangles. The measurement to use is the finished short side (fss) of the triangle plus 1/2" (1/4" seam allowance on each side). Place the BIAS SQUARE ™ ruler on top of the fabric triangle as shown. The points of the fabric triangle will stick out 3/8". Trim them off with the rotary cutter. The trim line is perpendicular to the short side of the triangles.

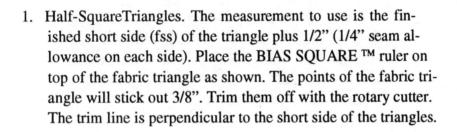

Half-Square Triangles

2. Quarter-Square Triangles. Draw a line with a marking pen on a cutting ruler that begins at the 3/8" mark and extends at a 45° angle across the ruler. For the first cut, place the ruler on top of the fabric triangle with the bottom edge of the ruler lined up with the long side of the triangle and the 45° drawn angle lined up with the 45° angle of the triangle. Trim off the 3/8" that sticks out. The trim line is perpendicular to the long side of the triangle. To trim the other triangle point, turn the ruler over and repeat.

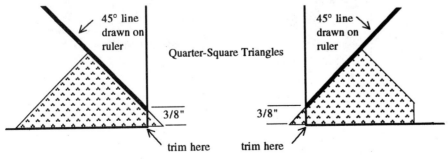

Quarter-Square Triangles

GUIDE TO ROTARY CUTTING

STRIP PIECING

In patchwork, every edge must be cut and every seam must be sewn. Sometimes the order of these two operations can be changed to save time or to get more accurate results. In strip piecing, fabric strips are cut either on the straight grain or bias, and sewn together in units called strata. The strata are then cut into shorter portions and then sewn together to form simple designs.

STRAIGHT STRIP PIECING

Straight strip piecing is a great time-saver for checkerboard-type designs like Four Patches, Ninepatches and Irish Chains, and for piecing portions of designs that have regular repetitions of squares and rectangles.

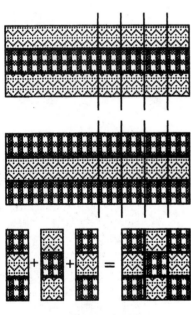

Ninepatch units

1. Cut strips from the length of the fabric when possible because it is easier to keep on grain. When it is necessary to use the cross grain, straighten the fabric so strips will be cut as close to true grain as possible. To determine the width to cut strips, add a 1/4" seam allowance to each side of the finished dimension of the desired shape. For example, if the finished square will measure 2", cut 2 1/2" strips.

2. Sew strips together with 1/4" seams.

3. Press seam allowances toward the darker fabric, pressing from the top so the fabric won't pleat along the seam line. Usually, pressing to the dark will result in opposing seams at the points of matching. If the coloring of the strips doesn't work out that way, press for opposing seams instead of always to the dark.

4. Measure and make crosswise cuts with ruler and rotary cutter.

5. Join the cut units with 1/4" seams to make the desired design.

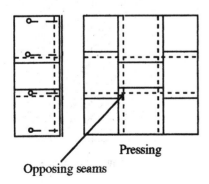

Opposing seams   Pressing

GUIDE TO ROTARY CUTTING

Bias Strip Piecing

In this strip-piecing method, strips are cut on the bias rather than on the straight grain, and the cut shapes are triangles rather than squares and rectangles. Because the seams are sewn on the bias edges, the resulting two-triangle units have the straight grain along the outer edge. It is an accurate method for piecing squares made of half-square and quarter-square triangles and can be adapted to other shapes as well.

Cutting Bias Strips. To avoid struggling with a lot of yardage, cut the fabric that will be cut into bias strips into manageable pieces. I prefer squares or rectangles that measure 14" to 22" and fit nicely on my cutting mat.

Cut size of Bias Square	Strip Width
1" to 1 1/4"	1 1/2"
1 1/2" to 1 3/4"	2"
2"	2"
2 1/4"	2 1/4"
2 1/2"	2 1/2"
2 3/4"	2 3/4"
3"	2 3/4"
3 1/4"	3"
3 1/2"	3 1/4"
3 3/4"	3 1/2"
3 7/8"	3 5/8"
4"	3 3/4"

1. Cut two large pieces of contrasting fabric and place right sides together. Both layers will be cut at the same time.

2. Align the 45° angle marking of the BIAS SQUARE™ along a straight grain edge of the fabric and use a long ruler to make the first cut. If your fabric is cut in squares, simply make the first cut diagonally corner to corner.

3. Measure the desired width of the strip from the first cut and cut again. Continue until the whole square has been cut into bias strips. The width of the bias strips will depend on the desired width of the two-triangle square you are cutting. The table to the left shows strip widths for different sized squares.

4. Pick up matching strips and sew the strips together on the long bias edge, using 1/4" seam allowance. Press seams open for squares cut 1 3/4" or smaller. Press seams toward darker fabric for larger squares. Then sew two or three bias strip pairs together as shown and press.

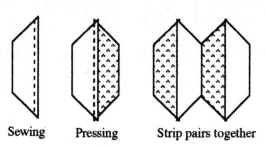

Sewing   Pressing   Strip pairs together

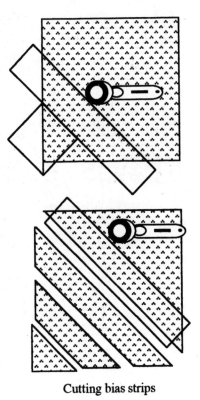

Cutting bias strips

GUIDE TO ROTARY CUTTING

Half Square
Triangle Unit

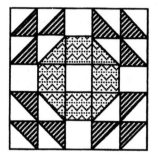

Half-square Triangle Units. These pieced units are called bias squares after the ruler used to cut them, and are used in Sawtooth designs, Feathered Stars, Oceans Waves and many others.

A variety of bias square fabric combinations can be obtained by sewing strips of different colors together. For instance, if you only want bias squares in a two-color combination, alternate light and dark strips. If you want a scrappy look with many different fabric combinations in your bias square units, cut strips of many different fabrics and sew them together as shown.

1. Using a BIAS SQUARE™ and rotary cutter, begin at the lowest points as shown and cut bias squares slightly larger than the desired cut size. If you have joined two or more sets of strips, cut squares from alternate rows working across the strips. After cutting the first set of squares, go back and cut from the skipped rows.

2. True up each cut square to the exact desired dimension.

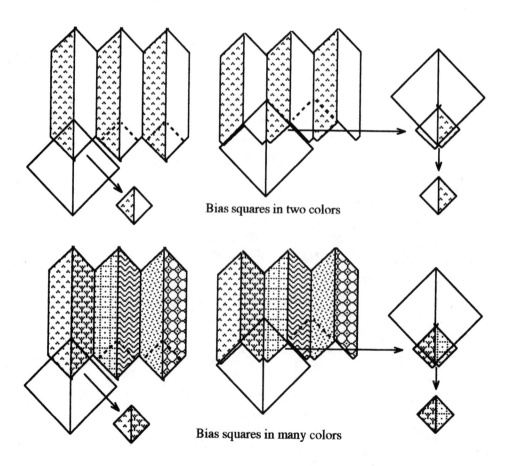

Bias squares in two colors

Bias squares in many colors

GUIDE TO ROTARY CUTTING

WASTE TRIANGLES

When cutting bias squares as described here, there will be triangular scraps or "waste triangles" left over from the outside edge strips. These pieces need not be wasted as they are a perfect size for single half square triangles in the same scale as the cut bias squares.

To trim the waste triangles to the proper size, follow these steps. Though this works with any size triangle, the example shown is a 2" finished half-square triangle.

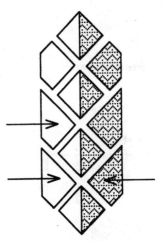
Waste triangles

1. Use the BIAS SQUARE™ and rotary cutter to true up the 90° corner of the waste triangle.

2. To trim the points of the triangle for easy matching, set the BIAS SQUARE™ at the 2 1/2" mark on the fabric piece as shown. Trim off the points that stick out. For other sizes of waste triangles, use the dimension that is the same size as the bias squares cut for your design.

3. To cut the diagonal of the triangle, place the BIAS SQUARE™ with the 45° angle line on the short side of the waste triangle. Look carefully at the illustration: position the trimmed point of the fabric triangle so the corner meets the 45° angle line where the two 1/4" markings meet. The trimmed line then continues out to meet the outside of the ruler at the 1/2" mark. Trim the diagonal with the rotary cutter.

1.
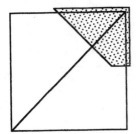

2.
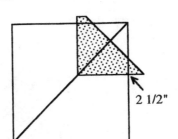
2 1/2"

3.
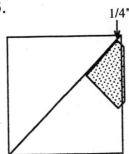
1/4"

19

GUIDE TO ROTARY CUTTING

Quarter Square
Triangle Unit

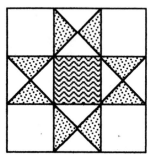

Quarter-Square Triangle Units. These pieced units are constructed from bias squares, and are found in many designs including the Ohio or Variable Star.

The example given is for a quarter-square triangle unit with a finished size of 3".

1. Cut bias strips 3 5/8" wide. (See table on page 17 for other strip widths.) Sew strips together with 1/4" seams. Press seams to the dark. Sew strip pairs together.

2. Using the BIAS SQUARE™, cut 3 7/8" bias squares. Squares are cut 7/8" larger than the finished quarter-square triangle unit.

3. Match pairs of cut bias squares, nesting opposing seams (fig. 1). Cut pairs diagonally and sew resulting triangle pairs with 1/4" seams (fig. 2) to complete unit. Press seams to one side (fig.3). Trim triangles points if desired.

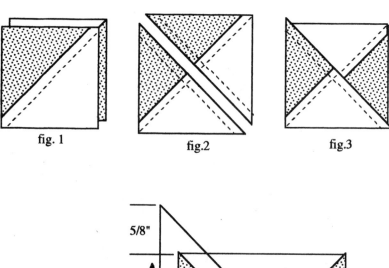

fig. 1   fig.2   fig.3

Finished dimension of Quarter Square Triangle Unit

1/4" + 5/8" = 7/8"

Add 7/8"

Detail of Cutting Guide on page 5.

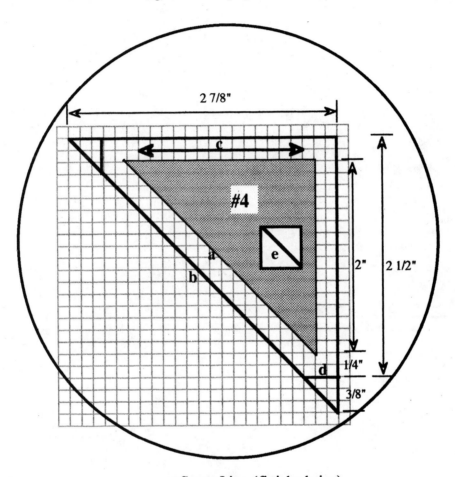

a. Seam Line (finished size)
b. Cutting Line (cut size)
c. Grain Line
d. Trim Line
e. Quick Cut Icon

GUIDE TO ROTARY CUTTING

Eliza's Ninepatch

Use the Cutting Guide below to cut pieces for this 6" patchwork block. See pages 4 and 5 for a discussion of Cutting Guides and how to use them. See pages 8 and 9 for how to cut squares (templates #1 and #3); and page 11, for how to cut quarter-square triangles (template #2). The center of this block is a ninepatch that could easily be strip pieced (see page 16).

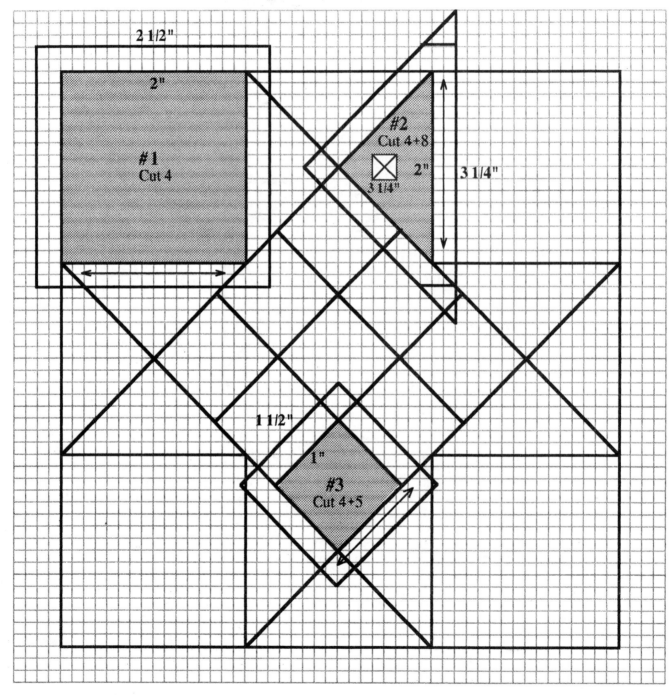